STEAM IN THE KITCHEN

SORTING
IN THE KITCHEN

THEIA LAKE

PowerKiDS press

Published in 2024 by The Rosen Publishing Group, Inc.
2544 Clinton Street, Buffalo, NY 14224

Copyright © 2024 by The Rosen Publishing Group, Inc.

All rights reserved. No part of this book may be reproduced in any form without permission in writing from the publisher, except by a reviewer.

First Edition

Editor: Theresa Emminizer
Book Design: Rachel Rising

Photo Credits: Cover, p.1 J2R/Shutterstock.com; pp. 4,6,8,10,12,14,16,18,20 porcelaniq/Shutterstock.com; p. 5 fizkes/Shutterstock.com; p. 7 CHALERMCHAI99/Shutterstock.com; p. 9 LanaSweet/Shutterstock.com; p. 11 Jochen Schoenfeld/Shutterstock.com; p. 13 Serg64/Shutterstock.com; p. 15 Sentelia/Shutterstock.com; p. 17 Valentina_G/Shutterstock.com; p. 19 pikselstock/Shutterstock.com; p. 21 Odua Images/Shutterstock.com.

Cataloging-in-Publication Data

Names: Lake, Theia.
Title: Sorting in the kitchen / Theia Lake.
Description: New York : Powerkids Press, 2024. | Series: STEAM in the kitchen | Includes glossary and index.
Identifiers: ISBN 9781499443721 (pbk.) | ISBN 9781499443738 (library bound) | ISBN 9781499443745 (ebook)
Subjects: LCSH: Set theory–Juvenile literature. | Food–Juvenile literature. | Kitchen utensils–Juvenile literature. | Kitchens–Juvenile literature.
Classification: LCC QA248.L35 2024 | DDC 428.1–dc23

Manufactured in the United States of America

Some of the images in this book illustrate individuals who are models. The depictions do not imply actual situations or events.

CPSIA Compliance Information: Batch #CWPK24. For Further Information contact Rosen Publishing at 1-800-237-9932.

Find us on

CONTENTS

Getting Started 4
Sort by Flavor 6
Sort by Shape 8
Sort by Color 10
Sort by Size 12
Try It Out! 14
Sort by Use 16
More Ways to Sort 18
Time to Eat! 20
Glossary 22
For More Information 23
Index .24

Getting Started

Sorting is arranging, or placing, things together based on how similar, or alike, they are. There are lots of ways to practice sorting in the kitchen. You can sort dishes, kitchen tools, or foods. Just be sure to ask your parent or caregiver first. Let's get started!

5

Sort by Flavor

Gather some different kinds of fruit. Then grab two big bowls. Sort your fruit by flavor, or taste. Place the sweet fruits into one bowl. Place the sour fruits into the other bowl. Here are some ideas:

Sweet
- bananas
- strawberries
- blueberries

Sour
- lemons
- oranges
- some grapes

← sweet

← sour

Sort by Shape

Gather some vegetables and place them on the table with your fruit. Now try sorting your food by shape. Place the circle-shaped foods on one side of a table. Place the oval-shaped foods on the other.

Circles Shapes
- apples
- onions
- oranges

Oval Shapes
- avocados
- brussels sprouts
- watermelons

oval

circle

9

Sort by Color

Next, try sorting your fruits and vegetables by color. How many red foods do you have? How many green foods? Here are some ideas to get you started:

Orange Foods
- carrots
- oranges

Green Foods
- celery
- green grapes
- spinach

Red Foods
- apples
- red bell peppers
- strawberries
- tomatoes

red orange green

Sort by Size

Now try sorting your food by size! For this activity, you could grab a ruler or measuring tape. Practice measuring while you sort by size! Here are some ideas:

Big
- cabbage
- grapefruit
- watermelon

Small
- blueberries
- grapes
- raspberries

Medium
- apples
- carrots
- lemons
- oranges

small →

medium ↓

← big

13

Try It Out!

Let's try sorting some different foods. Look in your cupboards and take out a few dry foods, such as pasta, rice, or beans. Lay them out on a table. How could you sort these foods? Should you sort them by flavor, size, shape, or color?

15

Sort by Use

You can also practice sorting with kitchen tools. Gather an **assortment** of tools and place them on a table. Sort the tools by their purpose, or how they're used. Here are some ideas:

Used for Measuring
- measuring spoons
- measuring cups
- kitchen **scale**

Used for Stirring
- whisk
- rubber spatula
- spoon

More Ways to Sort

You can also practice sorting while doing everyday **tasks**! You can sort dirty dishes while you load them into the dishwasher. Plates go in one spot, **utensils** in another. You can sort the clean dishes by size and use as you put them away in the cupboard.

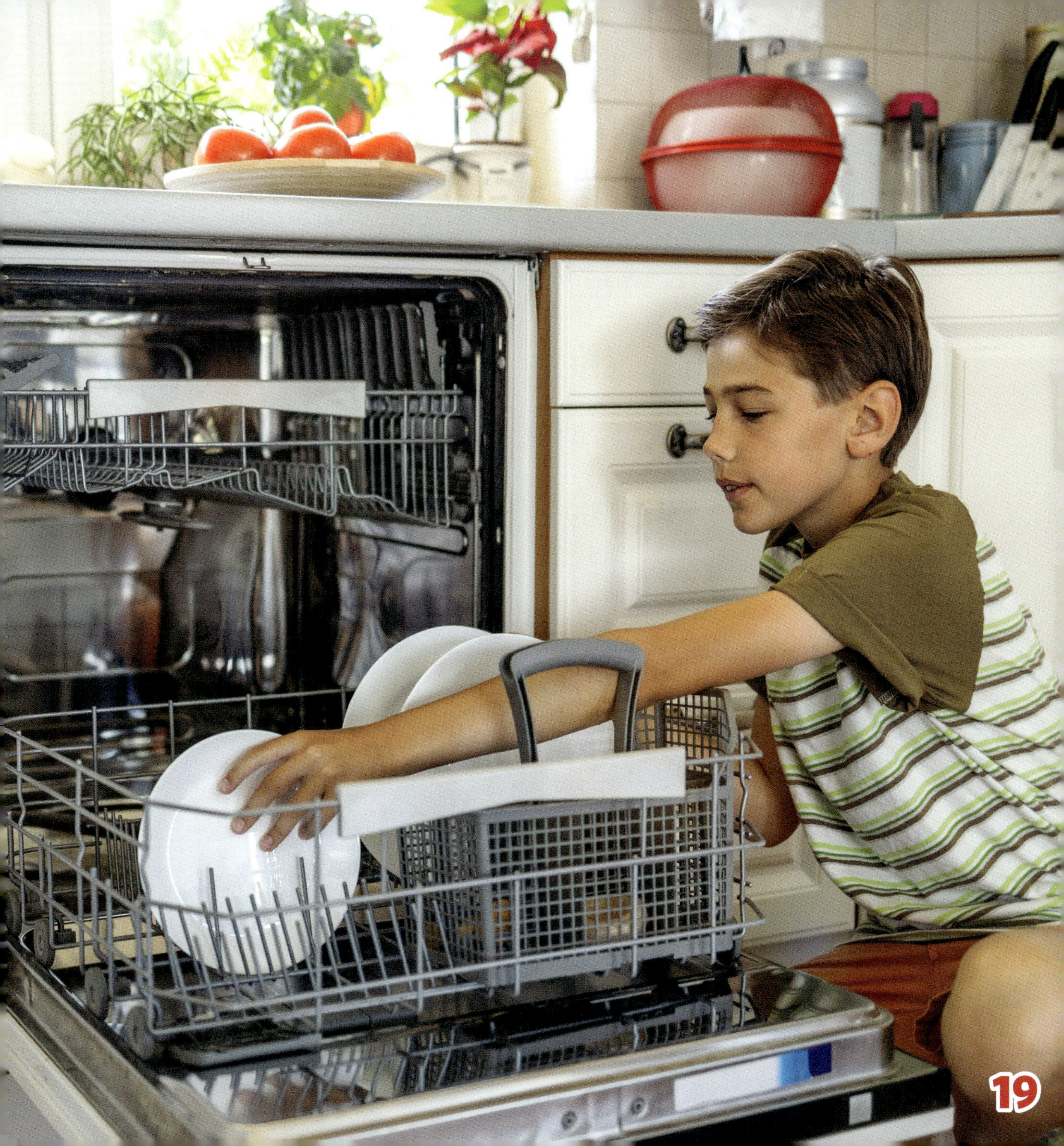

Time to Eat!

Did all that sorting make you hungry? It's time to eat! Make yourself a meal. You can even practice sorting on your plate! Sort the vegetables into one spot, the **grains** into another, and the meat into another. What other ways can you sort your foods?

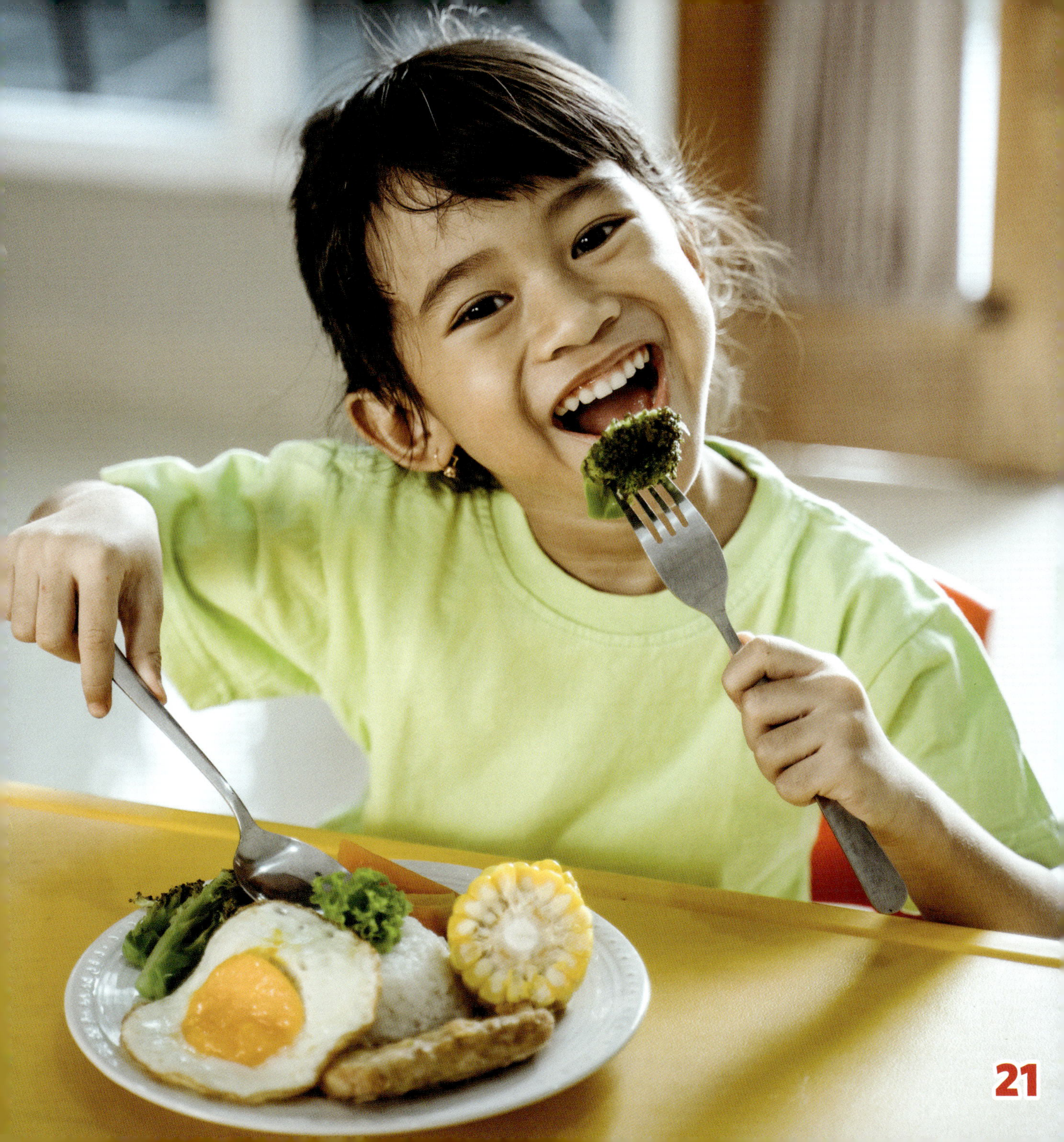

GLOSSARY

assortment: A group of different kinds of things.

grain: A food made with oats, wheat, rice, cornmeal, barley, or another cereal grain.

scale: A device, or machine, used for measuring weight.

task: A job, chore, or thing to do.

utensil: Cutlery or a tool used for eating, such as a fork, spoon, or knife.

whisk: A cooking tool that has curved wires and is used for stirring or beating ingredients, often eggs or whipped cream.

FOR MORE INFORMATION

BOOKS

Sorting for STEM Big Skills Workbook. Wilkinsburg, PA: Scholastic Early Learners, 2019.

Virr, Paul, and Lisa Regan. *Sorting Puzzles*. New York, NY: Windmill Books, 2020.

WEBSITES

National Geographic Kids
kids.nationalgeographic.com/games/action-adventure/article/recycle-roundup-new
Practice sorting while cleaning up and recycling!

PBS Kids
pbskids.org/sid/games/sorting-box
Practice sorting with this fun game!

Publisher's note to educators and parents: Our editors have carefully reviewed these websites to ensure that they are suitable for students. Many websites change frequently, however, and we cannot guarantee that a site's future contents will continue to meet our high standards of quality and educational value. Be advised that students should be closely supervised whenever they access the internet.

INDEX

C
circle, 8, 9

D
dishes, 4, 18

F
fruit, 6, 8, 10, 12

O
oval, 8, 9

R
ruler, 12

S
sour, 6, 7
sweet, 6, 7

T
tools, 4, 16

V
vegetables, 8, 10